Learning Points

89 Activities and Actions for Coaching Call Center CSRs

Peter R. Garber

HRD Press, Inc. • Amherst • Massachusetts

Copyright © 2007, Peter R. Garber

The materials that appear in this book, other than those quoted from prior sources, may be reproduced for educational/training activities. There is no requirement to obtain special permission for such uses. We do, however, ask that the following statement appear on all reproductions.

> Reproduced from *Learning Points: 89 Activities and Actions for Coaching Call Center CSRs* by Peter Garber, Amherst, MA: HRD Press, 2007.

This permission statement is limited to reproduction of materials for educational or training events. Systematic or large-scale reproduction or distribution—or inclusion of items in publications for sale—may be carried out only with prior written permission from the publisher.

Published by: HRD Press, Inc.
22 Amherst Road
Amherst, Massachusetts 01002
1-800-822-2801 (U.S. and Canada)
1-413-253-3488
1-413-253-3490 (fax)
http://www.hrdpress.com

ISBN-10: 0-87425-388-8
ISBN-13: 978-0-87425-388-7

Cover design by Eileen Klockars
Production services by Anctil Virtual Office
Editorial services by Sally M. Farnham

Table of Contents

Introduction vii

Part 1: The Changing Role of the Supervisor 1

    Learning Point 1:   Check Your Understanding 3
    Learning Point 2:   The Changing Role of the Supervisor 6
    Learning Point 3:   Changing Supervisor's Role Exercise 7
    Learning Point 4:   From Boss to Coach 8
    Learning Point 5:   CSR Coaching Tips 9
    Learning Point 6:   Coach's Changing Responsibilities 10
    Learning Point 7:   Boss-to-Coach Exercise 11
    Learning Point 8:   Your Best Coach 13
    Learning Point 9:   Building Trust 14
    Learning Point 10: Coaching Characteristics 15
    Learning Point 11: Check Your Understanding 16

Part 2: Motivating Others 19

    Learning Point 12: Recognition 21
    Learning Point 13: Which is the Best Way to Recognize? 22
    Learning Point 14: Recognizing CSRs 23
    Learning Point 15: Recognition Tips 25
    Learning Point 16: Reinforcement 26
    Learning Point 17: Reinforcement Exercise 27
    Learning Point 18: Keeping Score 28
    Learning Point 19: Scoreboard Tips 29
    Learning Point 20: Scoreboard Exercise 30
    Learning Point 21: Scoreboard Facts 31
    Learning Point 22: Check Your Understanding 32

Part 3: Creating Effective Communications 35

    Learning Point 23: Developing Better Listening Skills 37
    Learning Point 24: How We Send and Receive Messages 38
    Learning Point 25: What Happens Over the Telephone? 39
    Learning Point 26: Communications Exercise 40
    Learning Point 27: CSR Listening Tips 41
    Learning Point 28: Listening Facts 42
    Learning Point 29: Communicating is an Art 43
    Learning Point 30: No Absolutes in Communications 44
    Learning Point 31: Different Meanings for Different People 45

Learning Point 32: Make Yourself Clear 46
Learning Point 33: Be Aware of Communications Obstacles 47
Learning Point 34: Different Meanings 48
Learning Point 35: An Effective Communications Model 49
Learning Point 36: Feedback Sources 50
Learning Point 37: 360-Degree Feedback 51
Learning Point 38: Feedback is an Opportunity to Grow 52
Learning Point 39: Feedback Sources 53
Learning Point 40: Getting the Most from Feedback 54
Learning Point 41: Check Your Understanding 56

Part 4: Helping CSRs Work Together as a Team 59

Learning Point 42: Teamwork in Call Centers 61
Learning Point 43: Collaboration 62
Learning Point 44: Accountability 63
Learning Point 45: Standards 64
Learning Point 46: Synergy 65
Learning Point 47: Trust 66
Learning Point 48: Teamwork Tips 67
Learning Point 49: Teamwork Exercise 68
Learning Point 50: Characteristics of a Good Team Player 69
Learning Point 51: Team Facts 70
Learning Point 52: Work Teams 71
Learning Point 53: Work Teams on the Rise 72
Learning Point 54: Productive Work Teams 73
Learning Point 55: Supervisor's Teamwork Role 74
Learning Point 56: Creating Synergy 75
Learning Point 57: Leading Teams Exercise 76
Learning Point 58: Work Team Tips 77
Learning Point 59: Work Team Facts 78
Learning Point 60: Coaching is a Process 79
Learning Point 61: Coaches Don't Just Help People 80
Learning Point 62: Coaching Opportunities 81
Learning Point 63: Coaches Create Opportunities 82
Learning Point 64: Becoming a Role Model 83
Learning Point 65: Teamwork Tips 84
Learning Point 66: Teamwork Training Opportunities 85
Learning Point 67: Teamwork Training Exercise 86
Learning Point 68: Teamwork Facts 87
Learning Point 69: Check Your Understanding 88

Part 5: Resolving Conflicts Between CSRs 91

    Learning Point 70: Dealing with Upset CSRs 93
    Learning Point 71: Keep it on a Professional Level 94
    Learning Point 72: Keeping Everyone's Best Interest in Mind 95
    Learning Point 73: Letting People Vent 96
    Learning Point 74: Long-term Consequences of Conflict 97
    Learning Point 75: Tips for Dealing with Upset CSRs 98
    Learning Point 76: Dealing with Upset CSRs Exercise 99
    Learning Point 77: 7 Steps to Take to Help
                            Employees Resolve Problems 100
    Learning Point 78: Complaint Handling 101
    Learning Point 79: Understanding Complaints 102
    Learning Point 80: Investigating Complaints 103
    Learning Point 81: Complaint Handling Exercise 104
    Learning Point 82: Complaint Handling Tips 105
    Learning Point 83: Complaint Handling Facts 106
    Learning Point 84: Causes of Conflict 107
    Learning Point 85: Conflict Management Tips 108
    Learning Point 86: Conflict Exercise 109
    Learning Point 87: Conflict Resolution Matrix 110
    Learning Point 88: Resolving Conflict Review 111
    Learning Point 89: Check Your Understanding 113

Glossary of Terms 119

Introduction

Call centers are providing more convenient and efficient services to customers for virtually every business or organization today. Customers have learned to depend on these services to receive the products and services they purchase. But just how satisfied and happy customers are as a result of contacting a call center will determine the likelihood that they will call back again in the future. With the competition getting tougher and tougher each day, it is important when a customer calls a call center that he or she receives the highest quality service as possible. Ultimately it is the customer service representative, or CSR, on the other end of the telephone who will determine this level of customer satisfaction. Helping CSRs perform their jobs to the best of their abilities and in accordance with the customer service standards of the call center is a major responsibility of those who supervise these important positions.

Learning Points: 89 Activities and Actions for Coaching Call Center CSRs is designed to provide those who supervise CSRs in call centers to be better able to perform their jobs. The term *coach* is used rather than *supervisor* throughout this book to reflect the changing role of this position, but the terms are interchangeable. As a coach for CSRs, you play a critical role in the ultimate success of the call center. You are the person who can have the greatest overall impact or influence on how satisfied customers are with the service they receive when they call into the call center. CSRs look to you for the support and guidance they need to perform their jobs in the highest quality manner possible. Use the 89 Learning Points in this book to help those you coach to continuously improve their skills and provide the level of service that your customers expect and deserve.

Part 1:
The Changing Role of the Supervisor

Learning Point 1

Check Your Understanding

Take this brief questionnaire to check your understanding of the role of a supervisor or coach in a call-center environment.

1. The role of the supervisor in the workplace has changed very little over the past few years.
   a. True   b. False

2. Coaches only exist on sports teams.
   a. True   b. False

3. Motivation must always come from within.
   a. True   b. False

4. Keeping score at work might involve which of the following:
   a. Bulletin board notices about team accomplishments
   b. Articles in the company newsletter
   c. Graphs and charts
   d. Feedback from the supervisor or team coach
   e. All of the above

5. Effective communications is easy to achieve at work.
   a. True   b. False

6. You could receive valuable feedback on your performance at work from which of the following sources?
   a. Your boss or direct supervisor
   b. Your co-workers
   c. Your direct reports
   d. Suppliers and customers
   e. All of the above

7. Teamwork involves collaboration and cooperation.
   a. True   b. False

8. Work teams are always an appropriate way to address a problem at work.
   a. True   b. False

(continued)

Check Your Understanding (continued)

9. It is not a supervisor's or coach's role to get involved in conflicts between CSRs.
   a. True
   b. False

10. If a supervisor or coach ignores complaints from CSRs, which of the following would be a likely result?
    a. The CSR's complaint will go away.
    b. The CSR will resolve the problem him- or herself.
    c. The CSR will not be upset.
    d. The CSR will continue to harbor the complaint, and it may become an even bigger issue or problem.

Answers

1. The role of the supervisor in the workplace has changed very little over the past few years.
   a. True
   b. **False. Correct answer.** The role of the supervisor has changed over the years from one of telling employees what to do and ensuring that the work gets done to one that is more of a coach whose job it is to help people reach their greatest potential.

2. Coaches only exist on sports teams.
   a. True
   b. **False. Correct answer.** A coach can be someone in a supervisory role, a teacher, a member of the clergy, a parent, a friend, a co-worker, or anyone who helps you reach your goals.

3. Motivation must always come from within.
   a. True
   b. **False. Correct answer.** Supervisors or coaches also need to provide a motivating environment for employees.

4. Keeping score at work might involve which of the following:
   a. Bulletin board notices about team accomplishments
   b. Articles in the company newsletter
   c. Graphs and charts
   d. Feedback from the supervisor or team coach
   e. **All of the above. Correct answer.** Any form of feedback that tells employees how they are performing at work is a form of score keeping.

(continued)

Check Your Understanding (concluded)

5. Effective communications is easy to achieve at work.
   a. True
   b. **False. Correct answer.** Effective communication takes a great deal of hard work to achieve and maintain in any work place.

6. You could receive valuable feedback on your performance at work from which of the following sources?
   a. Your boss or direct supervisor
   b. Your co-workers
   c. Your direct reports
   d. Suppliers and customers
   e. **All of the above. Correct answer.** Each of these sources can provide valuable performance feedback to you.

7. Teamwork involves collaboration and cooperation.
   a. **True. Correct answer.** Everyone must collaborate to make teamwork successful at work.
   b. False

8. Work teams are always an appropriate way to address a problem at work.
   a. True
   b. **False. Correct answer.** There can be certain circumstances when establishing a work team is not appropriate or the best way to deal with a situation or problem.

9. It is not a supervisor's or coach's role to get involved in conflicts between CSRs.
   a. True
   b. **False. Correct answer.** There are times when a supervisor or coach must intervene and help resolve conflicts in the workplace.

10. If a supervisor or coach ignores complaints from CSRs which of the following would be a likely result?
    a. The CSR's complaint will go away.
    b. The CSR will resolve the problem him- or herself.
    c. The CSR will not be upset.
    d. **The CSR will continue to harbor the complaint, and it may become an even bigger issue or problem. Correct answer.** Complaints don't usually go away on their own. Supervisors or coaches need to take action to get complaints resolved one way or another.

Learning Point 2

The Changing Role of the Supervisor

It used to be that a supervisor's main job was to tell employees what to do and then to ensure that these tasks were completed. Supervisors were evaluated by how much work they were able to get their employees to complete.

There are both advantages and disadvantages to traditional methods of supervision. The advantages include everyone knowing what is expected of them on their jobs. The disadvantages include limiting employees' ability to have greater input into how the work is to be performed. There really isn't any right or wrong answer concerning the effectiveness of traditional supervision techniques. What is really most important is what works most effectively for a call center.

It is best to let the culture of the call center decide what style of supervision and leadership is most appropriate. This may further depend on a number of factors such as the type of business, leadership, and even the history of the organization.

Tradition itself often determines how a call center is managed and the way supervisors supervise. There is certainly nothing wrong with tradition. Tradition gives everyone something that they are accustomed to and are comfortable with. If it works for you, then there may not really be any reason to change the way you supervise.

Learning Point 3

Changing Supervisor's Role Exercise

The following continuum shows how decision making is implemented in both a traditionally supervised workplace and one that is nontraditional. Mark where you see how decisions are made in your call center and are most reflective of your style of supervision.

Traditional				**Nontraditional**
1	**2**	**3**	**4**	**5**
All decisions made by supervisors		Certain types of decisions shared with employees		Employees share in the decision-making process

Learning Point 4

From Boss to Coach

The role of the supervisor in many call centers is rapidly changing, which can be very confusing to supervisors trying to understand their position in today's work culture. The same supervisory skills that were needed and even expected in the past now may be viewed as ineffective or even inappropriate.

As the concepts of empowerment and teamwork become more and more commonplace in organizations, supervisors need to have a better understanding of the skills and characteristics that are expected of them.

Most organizations today are changing the culture of their work environment to one based more on the concepts of teamwork. This has been found to be a more enjoyable as well as productive way to work for everyone in the organization. This new way of working together creates many challenges for everyone on all levels of the organization, especially the supervisor.

The supervisor plays a major role in this change process. In fact, much of the success of this new working environment is dependent on this new role that the supervisor must play. Simply stated, the role of the supervisor today is changing from that of being the boss to one of becoming more of a coach. Instead of telling people what to do and then ensuring that the work is done, the effective supervisor today helps others learn to make decisions for themselves.

The roles of many positions in call centers today are rapidly changing to keep pace with the ways that organizations themselves are changing. Concepts such as empowerment, teamwork, quality improvement, and participative management are forever changing the roles and responsibilities of everyone's jobs. The skills needed in the future for supervisors to be successful are very different than they were in the past.

Learning Point 5

CSR Coaching Tips

- Think of ways in which you can better utilize the capabilities and potential of those you supervise.

- "Let go" of those responsibilities you presently have that could be done appropriately by those who work for you.

- Trust others with more information, particularly information they need to assume a more responsible role.

- Take a broader view of your role in the organization.

- Gain a clear understanding of what is now expected of you in the new management style in your organization.

- Suggest ways in which you can make more significant contributions to your organization.

- Think of your new role as a challenge, not a burden.

- Help others learn to accept and adapt to their new responsibilities in today's challenging workplace.

Learning Point 6

Coach's Changing Responsibilities

The following are some of the responsibilities of a coach in the new work environments of today:

- Mentor
- Teacher
- Mediator
- Director
- Spokesperson
- Communicator

- Resource provider
- Motivator
- Technical advisor
- Consultant
- Ombudsman
- Counselor

Learning Point 7

Boss-to-Coach Exercise

A coach helps employees reach their highest potential to perform their jobs and make decisions for themselves rather than just telling employees what to do. Review the actions in the left hand column of the table below that represent the ways that a boss might get the work done in a traditional work environment. These examples are not presented as being necessarily right or wrong, but simply descriptive of the role that a traditional supervisor might play. In the right hand column, list what you believe would be the supervisor's new role as a coach rather than a boss in a more traditionally managed call center. See suggested responses on the following page.

The Changing Role of the Supervisor

From Boss . . .	. . . to Coach
Assigns jobs	
Checks quality	
Gives performance evaluations	
Gives directions	
Disciplines	
Provides training	
Approves vacations	
Assigns overtime	
Creates reports	
Reports information to upper management	

(continued)

Boss-to-Coach Exercise (concluded)

The Changing Role of the Supervisor (possible responses)

From Boss . . .	. . . to Coach
Assigns jobs	Employees assign own jobs
Checks quality	Employees responsible for quality
Gives performance evaluations	Creates peer feedback
Gives directions	Has employees help one another
Disciplines	Employees deal with discipline issues
Provides training	Establishes peer training programs
Approves vacations	Team develops vacation schedules
Assigns overtime	Employees decide and schedule OT
Creates reports	Employees gather data and report
Reports information to upper management	Employees present information to upper management

Learning Point 8

Your Best Coach

What does it take to be a great coach? Why do some coaches have winning teams year after year, while other coaches produce a win only occasionally?

Who can be a coach? Coaches are not just in the sports world. There are all different kinds of coaches. A coach can be a teacher, a member of the clergy, a friend, a co-worker, a parent or grandparent, or even a supervisor.

To answer the above question, think about some of the great coaches you have heard of or have known. What were some of the factors that help them achieve the level of success they reached?

List the skills, traits, abilities, qualities, etc., that you think made these people great coaches:

_____
_____
_____
_____
_____
_____
_____

Which of these characteristics do you think would be effective in dealing with the people you supervise at your call center and why?

_____
_____
_____
_____
_____
_____
_____

Learning Point 9

Building Trust

What one word would you use to describe the most important factor that should exist between a supervisor and the employees who report to him or her? The answer is **TRUST**.

The supervisor's new role of becoming more of a coach than a boss involves trusting others to make the same decisions that he or she would have made under the same or similar situations. Instead of telling employees what to do, a coach will help employees decide what needs to be done. Instead of checking up on employees to ensure that the work is getting done properly, a coach will make them more accountable for their own work. And instead of giving employees limited information, a coach will share as much information as possible with employees.

Learning Point 10

Coaching Characteristics

The following are some characteristics of an effective coach:

- Sets goals
- Recognizes and praises progress
- Explains decisions
- Solicits ideas
- Provides support
- Provides resources
- Listens
- Facilitates problem solving
- Accepts suggestions and ideas
- Gives credit
- Reinforces
- Provides assurance

Learning Point 11

Check Your Understanding

1. Which of the following best describes the traditional role of a supervisor?
    a. Telling employees what to do and then ensuring that these tasks are completed
    b. Getting their employees to produce as much work as possible
    c. Making all the decisions in the workplace
    d. All of the above

2. The supervisor's new role is more like which of the following?
    a. The supervisor needs to take a more controlling role in the new work environments of teamwork.
    b. The supervisor needs to play more the role of a coach who enables others to reach their greatest potential on the job.

3. Which of the following is *not* true concerning the new role of the supervisor?
    a. Supervisors need to make employees more accountable and responsible for their jobs.
    b. Many of the skills that supervisors might have used in the past might not be appropriate or effective in today's changing workplace.
    c. The importance of the role of the supervisor will be less in those workplaces moving toward the concepts of teamwork.

4. Great coaches only exist in sports.
    a. True
    b. False

(continued)

Check Your Understanding (continued)

5. Which of the following are some characteristics of an effective coach?
    a. Sets goals
    b. Recognizes and praises progress
    c. Explains decisions
    d. Solicits ideas
    e. Provides support
    f. Provides resources
    g. Listens
    h. Facilitates problem solving
    i. Accepts suggestions and ideas
    j. Gives credit
    k. Reinforces
    l. Provides assurances
    m. All of the above

Answers

1. Which of the following best describes the traditional role of a supervisor?
    a. Telling employees what to do and then ensuring that these tasks are completed
    b. Getting their employees to produce as much work as possible
    c. Making all the decisions in the workplace
    d. All of the above

2. The supervisor's new role is more like which of the following?
    a. The supervisor needs to take a more controlling role in the new work environments of teamwork.
    b. The supervisor needs to play more the role of a coach who enables others to reach their greatest potential on the job.

(continued)

Check Your Understanding (concluded)

3. Which of the following is *not* true concerning the new role of the supervisor?
    a. Supervisors need to make employees more accountable and responsible for their jobs.
    b. Many of the skills that supervisors might have used in the past might not be appropriate or effective in today's changing workplace.
    c. The importance of the role of the supervisor will be less in those workplaces moving toward the concepts of teamwork.

4. Great coaches only exist in sports.
    a. True
    b. False. Correct answer. Great coaches exist in almost all aspects of life. A coach can be a teacher, parent, friend, member of the clergy, or any significant person in your life.

5. Which of the following are some characteristics of an effective coach?
    a. Sets goals
    b. Recognizes and praises progress
    c. Explains decisions
    d. Solicits ideas
    e. Provides support
    f. Provides resources
    g. Listens
    h. Facilitates problem solving
    i. Accepts suggestions and ideas
    j. Gives credit
    k. Reinforces
    l. Provides assurances
    m. All of the above

*Part 2:*
Motivating Others

Learning Point 12

Recognition

Everyone needs recognition. Recognition can be one of the most important motivators that employees can experience. Without recognition, employees feel that all of their hard work and efforts to do a good job are not being appreciated.

1. **Vary recognition.** A supervisor can give recognition to CSRs in a variety of ways. Sometimes the more variety the recognition, the better. Employees appreciate it when their supervisor goes out of his or her way to recognize their work and accomplishments.

2. **Recognition doesn't have to be expensive.** The most effective forms of recognition are often the least expensive. It really doesn't have to cost any money to simply acknowledge a CSR's good work in the call center. Just telling the individual that his or her efforts are appreciated can be a very important form of recognition.

3. **Informal recognition.** Informal recognition includes all the ways of acknowledging employees' efforts and hard work that are simple to do and don't have to cost a lot of money. For instance, informal recognition can include simply saying "Thank you" to employees. It can be a card, a cup of coffee, a brief mention in the morning meeting, a note on the bulletin board, a note sent home, or any other way in which you can provide this informal recognition. The most important thing in providing informal recognition is that it be presented in a sincere and appreciative manner. When given in this way, it can be one of the most motivating forms of recognition you can provide to your employees.

4. **Formal recognition.** Recognition can also be formal. Formal recognition includes those types of recognition that might be considered "official." Programs such as employee of the month awards or other types of official awards would fall into this category. Formal recognition is also an important part of your recognition program for your employees in the Call Center.

Learning Point 13

Which is the Best Way to Recognize?

Informal vs. Formal

There really is no right or wrong answer to which way is the best way to recognize your employees. Both informal and formal forms of recognition can be very important to your employees. People appreciate being officially recognized for their accomplishments. Formal recognition provides the acknowledgment that employees work so hard every day to achieve. But there might be limits to how many and how often this type of recognition can be provided. Informal recognition can be provided more frequently and to more employees. The best recognition programs include a balance of both formal and informal recognition to employees.

Learning Point 14

Recognizing CSRs

The following are a variety of ways in which you can recognize employees. Decide which of these suggestions would be considered informal and which would be considered formal methods of recognition by marking each as either "I" for informal or "F" for formal.

____ Special plaque recognizing employee's accomplishment
____ Award ceremony
____ Promotion
____ Invitation to lunch with the boss
____ A thank-you note
____ A cup of coffee
____ A raise in salary
____ A congratulatory handshake
____ Being told "Thank you"
____ A mention at the morning meeting
____ A new title
____ Special parking space for a month
____ Being named "employee of the month"
____ A letter sent to employee's home complimenting the accomplishment
____ A picture of the employee's team in the call center's newsletter

(continued)

Recognizing CSRs (concluded)

Answers

- _F_ Special plaque recognizing employee's accomplishment
- _F_ Award ceremony
- _F_ Promotion
- _I_ Invitation to lunch with the boss
- _I_ A thank-you note
- _I_ A cup of coffee
- _F_ A raise in salary
- _I_ A congratulatory handshake
- _I_ Being told "Thank you"
- _I_ A mention at the morning meeting
- _F_ A new title
- _I_ Special parking space for a month
- _F_ Being named "employee of the month"
- _F_ A letter sent to employee's home complimenting the accomplishment
- _F_ A picture of the employee's team in the call center's newsletter

Learning Point 15

Recognition Tips

- Informal recognition can often be even more meaningful to recipients than formal recognition if it is perceived to be sincere.

- Recognition given incorrectly can actually become a negative factor in an organization if it is not perceived to be fair, particularly by those who didn't receive the recognition.

- What might be considered positive recognition by one employee might not be to another. Make sure that you do not inadvertently give employees something that is actually punishing rather than positive in your recognition efforts.

Learning Point 16

Reinforcement

Why do people act the way they do?

Research shows that people act the way they do for a variety of reasons, but one factor that is often found to be most important is reinforcement. Simply stated, people do something for which they will be positively reinforced. If something is reinforcing to someone, he or she will be much more likely to repeat that behavior in the future. This is what is called positive reinforcement.

Positive Reinforcement—R+

Positive reinforcement or R+ is a term that means that the result of a particular behavior was positive and therefore reinforced or strengthened. The more times a behavior is positively reinforced, the more likely it will continue in the future.

Negative Reinforcement—R−

Negative reinforcement or R−, on the other hand, means that an unwanted action will be taken as a result of a behavior. This negative serves to diminish the likelihood of the behavior being repeated in the future. Most forms of punishment or discipline are types of negative reinforcement.

How Much R+ is Enough?

Studies show that the more positive reinforcement there is in a work environment, the more desired behaviors exist. In other words, if you create a work environment that provides positive reinforcement for good work, you will increase the amount of this desired performance. There really isn't any limit on the amount of R+ you can have, but you do need to be careful not to keep providing the same reinforcers all the time. People need a variety of reinforcers to keep interested and motivated toward achieving good results.

How Much R− is too Much?

Should you have any negative reinforcement in your workplace? The answer is "yes." You will always need to have negative consequences for not obeying the rules. But the ratio of positive reinforcement to negative reinforcement should be at least 4 to 1. In other words, for every negative reinforcer, you should have at least 4 positive reinforcers.

Learning Point 17

Reinforcement Exercise

The following are a variety of things that most people find reinforcing. Check those that you would like to see increased when working with CSRs in the call center.

- ❏ Sending a letter of commendation
- ❏ Asking for advice or opinion
- ❏ Giving verbal praise
- ❏ Letting CSRs report work results to upper management
- ❏ Increasing responsibility
- ❏ Granting greater decision-making authority
- ❏ Sending memos on positive employee performance
- ❏ Passing along compliments from others
- ❏ Giving employee choice of tasks or assignments
- ❏ Providing quick follow-up on questions or requests
- ❏ Posting employee's name on bulletin board, recognizing accomplishments
- ❏ Offering training
- ❏ Assigning special projects
- ❏ Saying "Thank you"
- ❏ Asking upper management to make a personal phone call to the employee

Offering:
- ❏ Preferred work schedule
- ❏ Job rotation
- ❏ Promotion
- ❏ Pay raise
- ❏ Flextime
- ❏ Free coffee
- ❏ Plaques, trophies, certificates
- ❏ Clothing with call center's name or logo
- ❏ Free lunch
- ❏ Dinner for two

Learning Point 18

Keeping Score

Everyone needs to know the score. Just like in sports, everyone needs to know what the score is in the game against the competition. Just think about what a sporting event would be like if no one kept score or communicated the results of the contest. Keeping score at work is just as important. If employees don't know how they are doing against the competition, they won't know if they need to improve their performance or not.

Keeping score at work can involve many things. It might be a profit and loss statement or the number of units sold of the company's product. It might entail the amount of services provided to customers or the public. It could be charge backs or customer complaints. Keeping this kind of score might include any one of a myriad of different performance indexes that might be readily available for you to use.

In sports, it is always clear who the competition is—it is the other team you are playing against. However, it isn't always so clear for most employees to determine who their competition is. Sometimes it might seem like the competition is another department in your organization or even your co-workers.

The real competition is other call centers in different organizations that are trying to take your customers and business away from you. Keeping accurate score of how well you are performing against this competition should tell you exactly how you measure against these outside forces.

Keep scoreboards current. In sports, scoreboards are constantly being updated to reflect the current performance of the teams competing against each other. Scoreboards leave no doubt about who is ahead in the game as well as how much improvement is needed by the team that is behind in the contest. This same principle needs to be true for scoreboards at work. They too must be kept current so that everyone can know how they are performing as a team and what needs to be done to be successful against the competition.

Learning Point 19

Scoreboard Tips

- Scoreboards communicate important information concerning how people are performing their jobs at work.

- The usefulness of scoreboards is not just limited to sports. Scoreboards are also useful in the workplace as well.

- Scoreboards need to be visible to everyone who has an interest and need to see them.

- Scoreboards need to be kept current.

- Keeping scoreboards updated needs to be the responsibility of those whose performance is being measured.

- The information that scoreboards provide needs to be useful and easily understood.

- Once established, scoreboards need to be continued as long as there is a need and use for the information they communicate.

Learning Point 20

Scoreboard Exercise

What are some ways that you can provide some type of scoreboard to employees in your workplace?

___

Possible suggestions for scoreboards:
- Posters
- Letters
- Faxes
- Signs
- Bulletin board notices
- E-mails
- Voice mail messages
- Banners
- Electronic signs
- Computers
- Closed-circuit TV systems

Learning Point 21

Scoreboard Facts

- Everyone will have a better understanding of how well they are performing as a team.

- Progress toward shared goals will be better understood.

- Problems can be identified earlier on in the process as a result of sharing this information.

- People will be more motivated if they understand the results of their efforts by understanding their team's score.

- Where the team stands in relation to the competition will be better understood.

- Everyone will be better able to focus on moving in the same direction toward achieving the goals of the team.

Learning Point 22

Check Your Understanding

1. Recognition can be provided in which of the following ways?
    a. Simple and complex
    b. Uncomfortable and comfortable
    c. Sincere and insincere
    d. Formal and informal

2. Recognition done incorrectly can actually be ineffective.
    a. True
    b. False

3. Which of the following is an example of providing positive reinforcement or R+ in the workplace?
    a. Punishing anyone who makes a mistake
    b. Denying CSRs promotions they have earned
    c. Ignoring the good performance of CSRs in a call center
    d. Providing rewards to CSRs for their hard work and accomplishments

4. It is important to employees to understand how their team at work is performing.
    a. True
    b. False

5. Scoreboards help team members support one another.
    a. True
    b. False

(continued)

Check Your Understanding (concluded)

Answers

1. Recognition can be provided in which of the following ways?
   a. Simple and complex
   b. Uncomfortable and comfortable
   c. Sincere and insincere
   d. **Formal and informal. Correct answer.** Recognition doesn't always have to be formal or expensive. Sometimes the most effective recognition is informal or given spur of the moment.

2. Recognition done incorrectly can actually be ineffective.
   a. **True. Correct answer.** Attempts to reinforce employees may have a negative effect if it isn't something that they feel positive about receiving.
   b. False

3. Which of the following is an example of providing positive reinforcement or R+ in the workplace?
   a. Punishing anyone who makes a mistake
   b. Denying CSRs promotions they have earned
   c. Ignoring the good performance of CSRs in a call center
   d. **Providing rewards to CSRs for their hard work and accomplishments. Correct answer.** Positive reinforcement involves rewarding the desired behaviors of those who work for you.

4. It is important to employees to understand how their team at work is performing.
   a. **True. Correct answer.** People need to know how well or poorly their team is performing in order to understand how to maintain or improve their performance.
   b. False

5. Scoreboards help team members support one another.
   a. **True. Correct answer.** By having a better understanding of how the team is performing, team members can also have a better idea of what they need to do to help and support one another.
   b. False

Part 3:
Creating Effective Communications

Learning Point 23

Developing Better Listening Skills

It is said that communications is the key to achieving all your goals. What this means is that communications is part of the solution to virtually every problem you might face in the call center. Listening is an important part of becoming an effective communicator. You might say that good listeners are made not born. Listening is a skill you must learn to do well. It doesn't always come easily or naturally to people.

Learning Point 24

How We Send and Receive Messages

We Listen with Our Eyes

Studies have shown that 55 percent of the messages we receive are from nonverbal behaviors of others. What this means is that things like body language have a tremendous influence on how we interpret communications from others.

Voice Inflections

In face-to-face communications, approximately 38 percent of the message we receive from others is based on their voice inflections or how something is said. The very same words can be interpreted completely differently depending on the speaker's voice inflections.

Words

Words are only 7 percent of the messages we receive.

Incredibly, the actual words we speak typically make up only 7 percent of the message that is actually received by other people. What this means is that it is truly not *what* we say, but *how* we say it that is really most important.

Learning Point 25

What Happens Over the Telephone?

Studies have also shown that something very interesting happens over the telephone. Obviously there aren't any nonverbal communications over the telephone, so what happens to that 55 percent of communications (see Learning Point 24)? Interestingly, it comes across as voice inflections. When a customer is on the telephone, about 88 percent of the message that they receive is from the CSR's voice inflections. Again, it is truly not *what* we say, but *how* we say it that is really most important, particularly when speaking to a customer on the telephone.

Learning Point 26

Communications Exercise

To demonstrate the impact of voice inflections on a message, repeat the following statement, "How can I help you today?" giving it the following interpretations that appear below based on your voice inflections. Have another person listen to you repeat each statement and respond to what meaning they believe is being sent by your voice inflections. Share what your intended meaning was compared to what is perceived by the listener. To simulate speaking on the telephone, turn your backs to each other so that no nonverbal messages are sent or received.

"How can I help you today?"

Intended meaning based on your voice inflections:
1. That you are having a bad day and don't really want to be bothered
2. That you are distracted or have something else on your mind
3. That you are mad about something
4. That you are not very confident that you will be able to help the customer
5. That you don't think you should be helping the customer, but rather someone else
6. That you are in a hurry to get off the phone
7. That you actually are very interested in helping the customer

Learning Point 27

CSR Listening Tips

- Don't interrupt the customer.

- Show concern for the customer's feelings.

- Avoid jumping to conclusions.

- Ask questions for clarification.

- Don't try to finish the customer's sentences.

- Don't be non-responsive.

- Have patience with customers.

- Don't think about your response instead of what the customer has to say.

- Restate or paraphrase some of the customer's statements to ensure understanding.

- Pay close attention to what the customer has to say and don't let your mind wander.

Learning Point 28

Listening Facts

- People who get the facts right are usually good listeners.

- Listening involves more than just hearing words—it also involves listening for the real intent of the message.

- Most listening distractions can be avoided if you focus entirely on what the customer has to say.

- Effective listening is not a passive experience—it is one that takes considerable effort, but is worth it in the end.

Learning Point 29

Communicating is an Art

Effective communications is more art than science. There are certain rules and procedures to follow when communicating with others, but there is no exact science as in other disciplines or fields of study. Yet communications is one of the most important skills that a supervisor or coach in a call center must possess.

Learning Point 30

No Absolutes in Communications

Just as there are no absolute right or wrong answers in communications, there are also thousands of variations of the common theme of effective communications. Each person has his or her own style of communication. What might work well for one person might not work at all for another.

Learning Point 31

Different Meanings for Different People

In a typical dictionary, an average of 28 separate meanings are listed for each of the 500 most often used words in the English language. No wonder effective communications at work can be so difficult to achieve!

Learning Point 32

Make Yourself Clear

It is very important when dealing with others in the workplace that you clearly communicate exactly what your expectations are. This is one of the most important traits of an effective supervisor as well as a communicator. Make sure that those who work for you understand exactly what it is that you want them to achieve.

Learning Point 33

Be Aware of Communications Obstacles

It is important to understand what many of the obstacles are to effective communications. Realize that people hear different things than what may have been intended. Check for other people's understanding of the messages you send to them by asking them to repeat what they understood back to you. Then if any misunderstanding exists, you can clarify or correct.

Learning Point 34

Different Meanings

To illustrate just how many different meanings one word can have, consider the following definitions for the word *fast*:

- A person is *fast* when he/she can run rapidly.
- But something is also *fast* when tied down and cannot be moved.
- And colors are *fast* when they do not run.
- One is *fast* when he/she moves with suspect company.
- But this is not quite the same thing as playing *fast* and loose.
- A racetrack is *fast* when it is in good running condition.
- A green on a golf course is *fast* when the ball moves easily and quickly.
- A friend is *fast* when he/she is loyal.
- A watch is *fast* when it is ahead of time.
- To be *fast* asleep is to be deep in sleep.
- To be *fast* by is to be near.
- To *fast* is to refrain from eating.
- A *fast* may be a period of non-eating or a ship's mooring line.
- Photographic film is *fast* when it is sensitive to light.
- A CSR is *fast* when he/she can effectively handle many calls each day.

Learning Point 35

An Effective Communications Model

There are five steps to effective communications between two people.

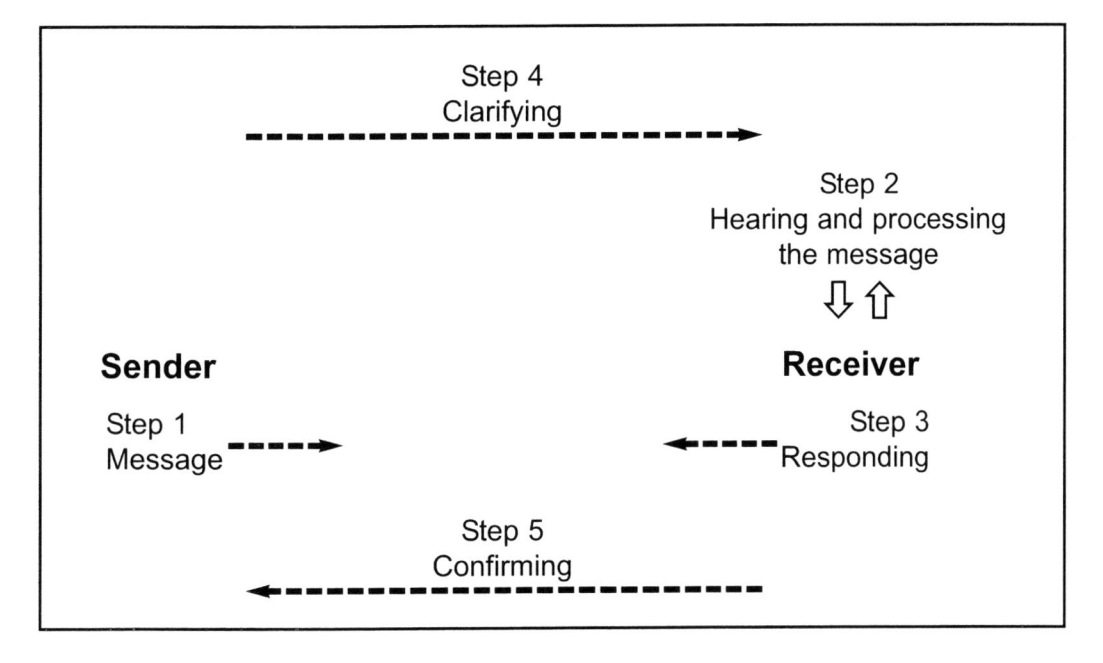

```
                        Step 4
                       Clarifying
        - - - - - - - - - - - - - - - - - - - ->

                                    Step 2
                              Hearing and processing
                                   the message
                                      ⇩ ⇧
    Sender                         Receiver

    Step 1                              Step 3
    Message  - - - >         < - - -   Responding

                        Step 5
                       Confirming
        <- - - - - - - - - - - - - - - - - - -
```

1. A person (the sender) sends a message to another person through his/her words.

2. The receiver of the message hears the words.

3. The receiver responds concerning what he/she heard the message to be.

4. The sender clarifies any misunderstanding that might have occurred in the transmitting of the message.

5. The receiver now confirms understanding of the message.

Try using these five steps when communicating with the CSRs who report to you. You will find that your communications are clearer and less frequently misunderstood.

Learning Point 36

Feedback Sources

As a supervisor or coach in a call center, you receive feedback on your performance from many different sources. Some of this feedback might be received as part of a formal performance appraisal process probably from your immediate supervisor or boss. But there are other sources of feedback as well.

Some of these other feedback sources might be from CSRs who report to you, customers, suppliers to the call center who provide products or services, and even your peers.

Learning Point 37

360-Degree Feedback

Alternative feedback sources may be formal, involving a learning tool typically called "360-degree feedback." This tool is designed to provide you with feedback from a variety of sources, not just from your immediate supervisor. Another way to receive performance feedback involves a person soliciting feedback from others with whom he/she works on their own, without the use of a formal feedback tool. Both of these methods are valuable because they can give you a broader perspective of how others are responding to you as you perform your job. Other sources of feedback may be more informal and be received as part of your daily contacts with others at work.

A 360-degree feedback tool typically includes input from a variety of sources concerning an individual's performance. Below is an example of how one of these feedback tools might be designed to provide a supervisor in a call center this input from a variety of sources.

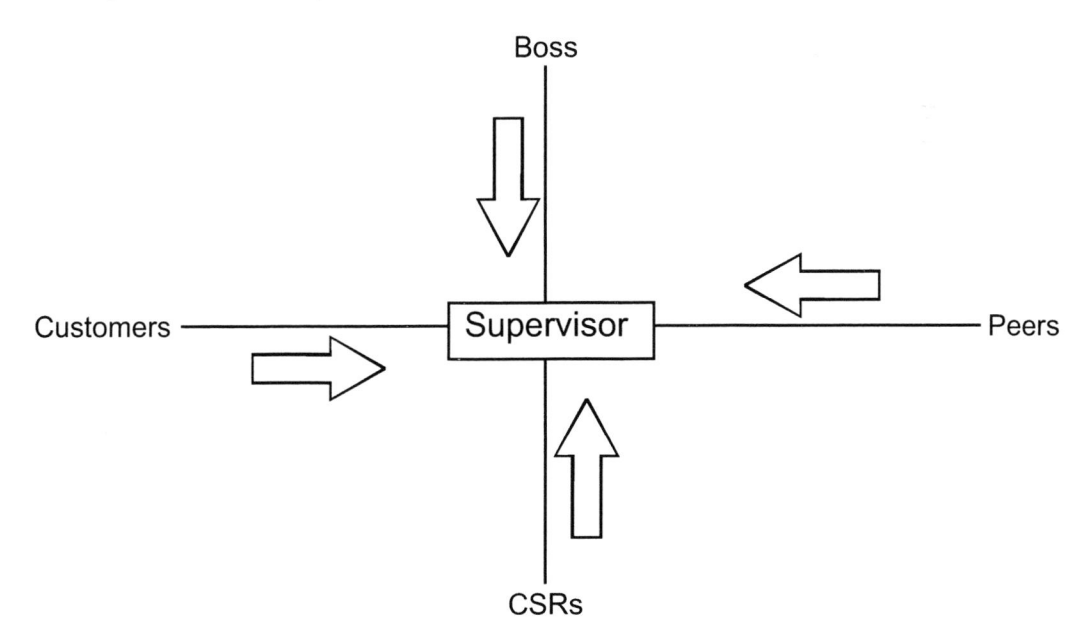

Learning Point 38

Feedback is an Opportunity to Grow

Receiving feedback about how we perform at work can be some of the most important information we can receive concerning our jobs. It can also be some of the most difficult and personal communications that we ever hear. How we perceive and receive feedback about ourselves can play a major role in its usefulness. We need to look at any chance to receive feedback about how we are performing our job as an opportunity to grow not only professionally, but personally.

Learning Point 39

Feedback Sources

There are many different ways in which supervisors in a call center can receive feedback from others about their performance. Most of these ways are somewhat subtle and easy to ignore or miss. However, this does not diminish their importance and potential value to you. The most important thing that you can do is to pay attention and even seek out these sources of feedback about yourself and your job performance. The following are just some of these various feedback sources:

- One-on-one contacts with others
- Comments others make about you
- The nonverbal behaviors of others who have contact with you
- The responsiveness of those who work with and for you
- The cooperation you receive from others
- The results that your work group achieves
- The number of questions others ask you about how to do their jobs
- The willingness of others to work with you or help you
- How your career is progressing

Learning Point 40

Getting the Most from Feedback

Receiving the most benefit from feedback requires extra effort. The following are possible suggestions for how supervisors can maximize the potential value and benefit of feedback from others they work with at a call center. Check each one as either "yes" or "no" if you think it will help you benefit from feedback or not.

Will this help you benefit from feedback?	Yes	No
You keep an open mind concerning the feedback you receive.		
You listen to only the feedback that you agree with from others.		
You ignore feedback from those with whom you don't get along.		
You ask for more information concerning feedback.		
You ask more people for feedback on your performance.		
You hold grudges against those who you think gave you poor feedback.		
Because you believe that receiving feedback once a year is enough, you discourage anyone giving you feedback except when your annual performance evaluation is due.		
You ask for follow-up feedback from those who gave it to you in the past to measure progress you might have made in improving your performance.		
You get angry and defensive concerning negative feedback that you receive from others.		
You pay attention to the subtle messages that people give you about your job performance.		

(continued)

Getting the Most from Feedback (concluded)

Answers

Will this help you benefit from feedback?	Yes	No
You keep an open mind concerning the feedback you receive.	X	
You listen to only the feedback that you agree with from others.		X
You ignore feedback from those with whom you don't get along.		X
You ask for more information concerning feedback.	X	
You ask more people for feedback on your performance.	X	
You hold grudges against those who you think gave you poor feedback.		X
Because you believe that receiving feedback once a year is enough, you discourage anyone giving you feedback except when your annual performance evaluation is due.		X
You ask for follow-up feedback from those who gave it to you in the past to measure progress you might have made in improving your performance.	X	
You get angry and defensive concerning negative feedback that you receive from others.		X
You pay attention to the subtle messages that people give you about your job performance.	X	

Learning Point 41

Check Your Understanding

Creating Effective Communications

1. Nonverbal communications make up what percent of the actual message we receive from other people?
   a. 5 percent
   b. 10 percent
   c. 40 percent
   d. 55 percent

2. What percent does voice inflections make up communications when talking to customers over the telephone?
   a. 88 percent
   b. 33 percent
   c. 66 percent
   d. 100 percent

3. Communications can best be described as which of the following?
   a. An exact science
   b. Exactly the same for everyone
   c. More of an art than a science
   d. Easy to master

4. Which of the following are the five steps in the correct order for effective communications to exist?
   a. Sender sends message, receiver hears it, receiver responds, sender clarifies, receiver confirms understanding
   b. Receiver hears message, sender sends message, receiver confirms understanding, sender clarifies, sender responds
   c. Sender responds, sender clarifies, receiver confirms understanding, sender sends message, receiver hears message
   d. Receiver confirms understanding, sender clarifies, receiver responds, receiver hears it, sender sends message

5. Which of the following may be sources of a 360-degree feedback system?
   a. Peers
   b. Supervisor
   c. Direct reports
   d. Customers
   e. Suppliers
   f. All of the above

(continued)

Check Your Understanding (continued)

Answers

1. Nonverbal communications make up what percent of the actual message we receive from other people?
   a. 5 percent
   b. 10 percent
   c. 40 percent
   d. 55 percent. Correct answer. Voice inflections make up 38 percent, and the actual words 7 percent.

2. What percent does voice inflections make up communications when talking to customers over the telephone?
   a. 88 percent. Correct answer. The 38 percent influence of voice inflections in face-to-face communications turns to 88 percent over the telephone.
   b. 33 percent
   c. 66 percent
   d. 100 percent

3. Communications can best be described as which of the following?
   a. An exact science
   b. Exactly the same for everyone
   c. More of an art than a science. Correct answer. Communications is more art than science, but there are certain principles that should be followed.
   d. Easy to master

4. Which of the following are the five steps in the correct order for effective communications to exist?
   a. Sender sends message, receiver hears it, receiver responds, sender clarifies, receiver confirms understanding. Correct answer and order of steps.
   b. Receiver hears message, sender sends message, receiver confirms understanding, sender clarifies, sender responds
   c. Sender responds, sender clarifies, receiver confirms understanding, sender sends message, receiver hears message
   d. Receiver confirms understanding, sender clarifies, receiver responds, receiver hears it, sender sends message

(continued)

Check Your Understanding (concluded)

5. Which of the following may be sources of a 360-degree feedback system?
   a. Peers
   b. Supervisor
   c. Direct reports
   d. Customers
   e. Suppliers
   f. **All of the above. Correct answer.**

Part 4:
Helping CSRs Work Together as a Team

Learning Point 42

Teamwork in Call Centers

Teambuilding is a very big factor at work today. The need for call center staff to work more effectively as a team is becoming a greater necessity every day. If call centers expect to remain competitive both today and in the future in our increasingly competitive world, they must continuously find better ways to work together as a team. Teamwork helps you reach your greatest potential both as an individual and as a team.

Learning Point 43

Collaboration

Collaboration means working together toward a common team goal, sharing both information and resources that enable each member to make the greatest contribution toward reaching this goal. As a supervisor, you must create a work environment in which collaboration can exist on your team.

Learning Point 44

Accountability

When it comes to teamwork, accountability means accepting personal responsibility for the performance and results of a particular function of the team. This might involve team projects or a particular function of the team. Team members must also accept accountability for their own jobs in the call center. For a team to function effectively, each member must accept this personal job accountability the same as they would any other responsibility in their life.

Learning Point 45

Standards

For any team, the standards to which each member is held must be clearly understood. Standards include the requirements, rules, regulations, and goals that must be met or exceeded by each member of the team.

Learning Point 46

Synergy

Synergy is the combining of human abilities and energies, resulting in the whole of a team being clearly greater than the sum of its members' talents and ability to reach their goals working as individuals. Synergy is the ultimate objective and benefit of teamwork. With teamwork, 2 + 2 > 4.

Learning Point 47

Trust

Trust comes from the support that team members give to one another in achieving the goals and objectives of the team. Trust is created when there is the feeling that this support will always be provided under any circumstances. As a supervisor or coach, creating this trust is one of the most important goals for creating a team environment.

Learning Point 48

Teamwork Tips

- Use the basic teamwork principles of collaboration, accountability, standards, synergy, and trust (see Learning Points 43–47) to build a stronger team in your call center.

- With your team, set goals in each of these areas.

- Measure your progress against these goals, involving team members in the process.

Learning Point 49

Teamwork Exercise

Think about ways in which the principles of teamwork described in this module can be applied to your team at the call center. In the appropriate column, describe examples of when these principles were used on your team, or future opportunities to use each teamwork principle.

Principle	Already Used	Use in the Future
Collaboration		
Accountability		
Standards		
Synergy		
Trust		

Learning Point 50

Characteristics of a Good Team Player

1. Openly shares feelings, opinions, thoughts, and perceptions about problems and conditions

2. When listening, attempts to hear and interpret communications from sender's point of view

3. Utilizes resources, ideas, and suggestions of other team members

4. Trusts and supports other team members and encourages their growth and development

5. Understands and is committed to team objectives

6. Acknowledges and works through conflict openly, by respecting and being tolerant of individual differences

7. Makes decisions based on information only, rather than being influenced by status or organizational role

8. Always strives for a win/win solution

9. Strives for consensus on a team decision

Learning Point 51

Team Facts

- Teams can accomplish more than individuals working independently.

- Collaboration can be one of your greatest natural resources available to you as a coach or supervisor.

- CSRs will be willing to accept greater responsibility and accountability on their jobs.

- It is your role as supervisor or coach to ensure that everyone understands what is expected of them as a member of the team and to provide the resources necessary for each CSR to meet these expectations.

Learning Point 52

Work Teams

Work teams are becoming more and more popular as a way to solve problems in organizations that often elude individuals working by themselves. A work team offers the opportunity for its team members to feel that they can make important contributions to the organization. A work team creates energy and enthusiasm to find better ways to do things that are beneficial to both its members and to the entire organization.

Learning Point 53

Work Teams on the Rise

The use of work teams has dramatically increased in many organizations during the past few years. Work teams consisting of members from all levels of the organization are finding better, more innovative solutions to problems that many times were previously unsolved.

Learning Point 54

Productive Work Teams

Participating on a work team can be a very productive and rewarding experience for everyone involved. Work teams can solve problems that elude individual efforts. Work teams can also create a stronger sense of teamwork in an organization.

Learning Point 55

Supervisor's Teamwork Role

As a supervisor, it is your role to help people work together as a team. Creating work teams to address certain problems can be an effective way to help employees reach their highest potential and make the greatest contributions.

Learning Point 56

Creating Synergy

Work teams combine the knowledge, experience, talents, intuitions, and creativity of each member in a focused effort. Synergy is combining these abilities of the members, resulting in the whole being greater than the sum of the parts of the team. With synergy, 2 + 2 = 5+

Learning Point 57

Leading Teams Exercise

The following are essential steps to leading successful work teams. Review each of the following and think about how you can ensure that you are providing these important factors for your work teams at the call center.

Leadership: A work team needs to have effective leadership to help it continuously move toward its goals.

Clear direction: Work teams must understand what is expected of them and what they should accomplish.

Resources: A work team must have the resources including time to be able to achieve its objectives.

Meeting place: A work team must have a place where members can meet and do their work.

Training: Work team members must be provided the training necessary to acquire the skills they need to be able to accomplish their goals.

Rewards: Work team members must be given rewards for their contributions in achieving goals.

Communications: Work teams must maintain communications with others in the organization concerning their projects.

Trust: Work team members must trust one another.

Creativity: Work teams must find new and creative ways to solve problems.

Learning Point 58

Work Team Tips

- Work teams in call centers can be used effectively to solve problems that have been difficult to find answers to in the past.

- Work teams need the clear direction and leadership to be successful.

- There may be certain situations that might not be appropriate for work teams to be assigned. A number of these situations will be reviewed in the next module.

Learning Point 59

Work Team Facts

There are certain times when work teams are and are not appropriate for a supervisor or coach to use in a call center. The following are examples of both instances:

When Work Teams are Most Appropriate

- When a problem requires a number of different viewpoints to be considered

- When members of the work team have an interest in the problem

- When the supervisor or coach is willing to at least consider what the work team recommends

- When members of the work team understand exactly what is expected of them

- When the team understands that working together on difficult issues is often a stressful but rewarding experience

When Work Teams are *Not* Appropriate

- When a decision about how to solve a problem has already been made

- When the problem involves highly confidential or sensitive information that cannot be shared with the team

- When the actions that the team would be required to take are not appropriate for the members based on their positions in the organization

- If the resources needed for the team to be successful are not available

- If there is no support for the work team from other parts of the organization, particularly upper management

- If the nature of the problem to be addressed is not within the expertise of the team members

Learning Point 60

Coaching is a Process

Coaching is an ongoing process of helping people reach their greatest potential, not only as a group but as individuals as well.

Learning Point 61

Coaches Don't Just Help People

Coaches don't just help people get their work or projects completed, although this can be a big part of their job. More importantly, coaches help others learn to help themselves.

Learning Point 62

Coaching Opportunities

One of the situations when coaching can be most effective is when employees have never had the opportunity to contribute their ideas and expertise to the organization or just need to upgrade their skills to meet the changing expectations of a call center today.

Learning Point 63

Coaches Create Opportunities

Coaches must often find opportunities for work teams and teamwork to exist in a workplace. This might take a little imagination or creativity and perhaps even risk. However, in the end, the results will be worth all the effort put into creating teamwork opportunities.

Learning Point 64

Becoming a Role Model

People seldom improve when they have no other role model but themselves to copy. They have to learn from example and role models. As a coach, you must set the example for the work behaviors that will be most conducive to and supportive of teamwork in your call center. If you don't do this, it really isn't fair to expect those who work for you to know how to grow into better team players.

Learning Point 65

Teamwork Tips

The following are ways for you as a coach to create a team environment:

- Identify growth opportunities for your employees.

- Seek learning experiences for both CSRs and teams.

- Create a work environment that is more of a partnership between you and those who report to you.

- Encourage contribution of ideas from both individuals and teams.

- Make employee development one of your top priorities.

Learning Point 66

Teamwork Training Opportunities

Help CSRs learn by providing appropriate training opportunities. Connect them with others in the organization who possess the skill, knowledge, and experience to teach them new things. Find courses or training programs that target the skills that would be helpful for teams and CSRs to develop.

Teach people how to learn from their successes as well as failures.

Learning Point 67

Teamwork Training Exercise

In many cases, people already have the skill or the knowledge they need, but lack the support or motivation to apply these skills. With this in mind, answer the following two questions to decide if you should focus on building new skills in your work team or if another strategy is needed.

1. Do the members of the team lack the skills they require to be successful, or are they just not applying them? What are some examples?

   _____
   _____
   _____
   _____
   _____
   _____
   _____
   _____
   _____
   _____

2. If they require new skills, are these skills clearly defined? In not, how can these skills be better defined?

   _____
   _____
   _____
   _____
   _____
   _____
   _____
   _____
   _____

Learning Point 68

Teamwork Facts

- "Experience by itself can be a poor teacher; it often gives the test before presenting the lesson. Experience is most useful when combined with the necessary training and guidance needed to learn how to perform a job to the person's greatest potential."

- "Give a man a fish and you feed him for a day. Teach a man to fish and you feed him for a lifetime."

- "The mountain gets bigger the closer you get to it." Teams understand problems better when they are part of the solution.

Learning Point 69

Check Your Understanding

Teambuilding

1. Which of the following describes the collaboration needed by CSRs to work effectively as a team?
   a. Only worries about his or her own job
   b. Does not share information with others that could help them do their job better
   c. Works together toward a common goal, and shares information as well as resources with one another
   d. Keeps resources all to him- or herself

2. Synergy means which of the following?
   a. The combined effort of the members is less than individuals
   b. The combining of human abilities and energies, resulting in the whole of a team being greater than the sum of its members' talents and ability to reach their goals working independently

3. Work teams are always appropriate in any situation at work.
   a. True
   b. False

4. Which of the following best describes the coach's role in leading teams at work?
   a. Ensuring that everyone understands what is expected of them as a member of the team and to provide the resources necessary for each CSR to meet these expectations
   b. Giving CSRs greater levels of responsibility as they become more knowledgeable of their jobs
   c. With greater responsibility, also giving CSRs greater accountability for their jobs
   d. Creating a collaborative work environment
   e. All of the above

(continued)

Check Your Understanding (continued)

5. Once you create a teamwork environment, your job as coach is completed.
   a. True. Teamwork is a natural process that once begun will continue to work well on its own.
   b. False. Coaching a teamwork environment is a continuous process of helping people reach their greatest potential, not only as a group but as individual contributors as well.

Answers

1. Which of the following describes the collaboration needed by CSRs to work effectively as a team?
   a. Only worries about his or her own job
   b. Does not share information with others that could help them do their job better
   c. **Works together toward a common goal, and shares information as well as resources with one another. Correct answer.**
   d. Keeps resources all to him- or herself

2. Synergy means which of the following?
   a. The combined effort of the members is less than individuals
   b. **The combining of human abilities and energies, resulting in the whole of a team being greater than the sum of its members' talents and ability to reach their goals working independently. Correct answer.**

3. Work teams are always appropriate in any situation at work.
   a. True
   b. **False. Correct answer.** There are circumstances when work teams might not be appropriate. A few examples of these circumstances may be if information is highly confidential, a decision has already been made, resources are not available, there is lack of management support, or the problem is not within the expertise of the team members.

(continued)

Check Your Understanding (concluded)

4. Which of the following best describes the coach's role in leading teams at work?
    a. Ensuring that everyone understands what is expected of them as a member of the team and to provide the resources necessary for each CSR to meet these expectations
    b. Giving CSRs greater levels of responsibility as they become more knowledgeable of their jobs
    c. With greater responsibility, also giving CSRs greater accountability for their jobs
    d. Creating a collaborative work environment
    e. **All of the above. Correct answer.**

5. Once you create a teamwork environment, your job as coach is completed.
    a. True. Teamwork is a natural process that once begun will continue to work well on its own.
    b. **False. Coaching a teamwork environment is a continuous process of helping people reach their greatest potential, not only as a group but as individual contributors as well. Correct answer.**

Part 5:
Resolving Conflicts Between CSRs

Learning Point 70

Dealing with Upset CSRs

Dealing with upset employees can be a very difficult part of a supervisor or coach's job. There are certain ways to more effectively deal with upset employees that a supervisor or coach needs to know. First and foremost, it is important that you deal with any upset CSR in a professional and responsible manner at all times.

Learning Point 71

Keep it on a Professional Level

As a supervisor and member of management, you should deal with an upset employee in a professional and responsible manner. This may at times be difficult to do, especially when employees become emotional and possibly confrontational. How you react to this type of situation will set the tone for the entire interaction between you and the upset employee. You need to model or demonstrate the behaviors that are acceptable and expected in this type of situation. By remaining professional and composed yourself will help the employee calm down and be better able to begin to express what is upsetting him or her. You may also find it appropriate in certain situations to remind an upset employee that no matter how he or she may be feeling at the moment, certain rules and expectations of behavior exist that still must be obeyed.

Learning Point 72

Keeping Everyone's Best Interest in Mind

As a supervisor or coach, you must keep in mind the best interest of the upset employee, the rest of the work group, and the entire organization when dealing with these types of situations. A supervisor or coach needs to deal with an upset employee on a number of different levels. There are certain needs that an employee has during these emotional moments of their lives and careers that can potentially have significant long-term consequences, both positive and negative.

Learning Point 73

Letting People Vent

As a supervisor, you can't effectively communicate with someone when he or she is upset—either emotional or angry. One of the most important things to do when dealing with upset employees is to let them get over at least some of their anger before trying to talk rationally with them about whatever it is that they are upset about.

Learning Point 74

Long-term Consequences of Conflict

How a supervisor deals with an employee when he or she is upset could be a determining factor in the rest of that person's career with the company and call center. If an employee feels that his or her supervisor isn't concerned with their problems, it can cause their working relationship to be negatively affected. In many ways, you need to look at dealing effectively with upset employees as an investment in your future working relationship.

Learning Point 75

Tips for Dealing with Upset CSRs

- Allow the person to vent his or her emotions; you really can't begin to understand the problem until this is done.

- Listen to what the person has to say without judgment, at least initially.

- Try to understand the circumstances that lead up to the employee's anger.

- Look at the situation from the employee's perspective.

- Suggest that you set another time to discuss the problem to give the person time to calm down.

Learning Point 76

Dealing with Upset CSRs Exercise

Think of a situation when you were faced with dealing with an upset employee.

How effectively did you deal with the situation?

_____
_____
_____
_____
_____
_____
_____

How could you have handled the situation better? What would you have done?

_____
_____
_____
_____
_____
_____
_____

What do you believe were the longer-term consequences of how you dealt with this situation?

_____
_____
_____
_____
_____
_____
_____

Learning Point 77

7 Steps to Take to Help Employees Resolve Problems

1. Assure the employee that you are interested in helping him or her deal with the problem.

2. Ask the employee for additional information or for clarification to ensure your understanding of the problem.

3. Provide support to the employee in resolving the problem, as appropriate to do so.

4. Have the employee assume primary responsibility as much as possible for resolving the problem, if work related.

5. Suggest other people or resources that may also be of assistance to the employee in resolving the problem.

6. Ask the employee what he or she wants to remain confidential concerning the problem and honor these requests.

7. Don't lower your performance standards even when an employee has personal problems that could impact his or her work. Find ways that the CSR can still maintain the work quality while dealing with his or her problem.

Learning Point 78

Complaint Handling

What is a complaint? A complaint is any condition in the call center that a CSR thinks or feels is unjust or inequitable. Even though a complaint may be seen from a very different perspective by other people, this does not diminish the importance of it to the person who originally had the concern or complaint.

Learning Point 79

Understanding Complaints

To really understand an employee's complaint, you need to listen to it with an open mind. Avoid jumping to conclusions or not letting the person finish his or her complaint before beginning to try to problem solve. Try to understand what the root cause of the complaint might be. Restate the complaint to the CSR's satisfaction to ensure understanding.

Learning Point 80

Investigating Complaints

Each CSR complaint must be investigated. The following steps should be taken when investigating CSR complaints:

1. **Get the facts.** Get all the facts, hear both sides of the story, and understand how the call center and company's policies may apply to the situation.

2. **Take action.** You need to take action to deal with the problem. Either refer the problem to the appropriate person or persons, or immediately deal with it yourself. Remember that the problem is the most important thing going on in this person's life at this time. Take it seriously. Avoid making a snap decision, but don't delay any longer than necessary. Give the person a decision as soon as possible, and explain why the decision was made.

3. **Follow up.** Follow up with the CSR about the complaint after a few days. If it was decided that an action should be taken, make sure that it has been implemented. Communicate any follow-up actions that should be taken to the appropriate person or persons at work.

Learning Point 81

Complaint Handling Exercise

1. What are employees' reactions to unsettled complaints? Are there costs to not being able to resolve employees' complaints? List below what you perceive to be some of these costs to unsettled or unresolved employee complaints:

   _____
   _____
   _____
   _____

   Possible responses:
   An employee can
   - Adjust to the situation
   - Take some kind of other action
   - Complain to others
   - Worry or brood

2. Is it good if a supervisor or coach doesn't hear any complaints? Why or why not?

   _____
   _____
   _____
   _____

3. Rewrite the following statement to make it a more effective way to deal with an upset employee who brings a problem to his or her supervisor:

   > "I don't care how you feel about the decision, you will just have to live with it like everyone else around here!"

   _____
   _____
   _____
   _____

   Suggested response:
   "I'm sorry that you are upset about how the decision is affecting you. You will have to follow the new rules, but what do you think could be done to make these changes go better for you?"

Learning Point 82

Complaint Handling Tips

The following are steps to take when receiving a complaint from a CSR:

1. Listen
2. Get the story straight
3. Investigate
4. Decide
5. Take action
6. Follow up

Learning Point 83

Complaint Handling Facts

- There will always be complaints.

- How you handle complaints can make a big difference in getting them resolved.

- The aim of complaint handling is to solve a complaint—*not to win it!*

Learning Point 84

Causes of Conflict

To understand how to resolve conflict, you must first have an understanding of how it is caused. Trying to resolve conflict without first understanding the cause would be like a doctor trying to treat a patient without first examining the person. The following are possible causes of conflict between CSRs:

1. **Miscommunications.** Miscommunication is a frequent cause of conflict in the workplace. It can be caused by many factors, including the CSR either not receiving a message or receiving only part of the message communicated. Or perhaps the message was delivered in a manner that may have been misinterpreted.

2. **Different interpretations.** Different interpretations of something can cause conflict among employees as well. The employee believes that adherence to rules, policies, or procedures should be carried out in one way, while the actual intent of the rule, policy, or procedure is something else entirely.

3. **Different values.** Employees often have differing values that lead to conflicts. Differing values may be caused by the employee having less interest or even regard than others for a specific task or duty and does not attach much importance to its value.

4. **Opposing goals.** Opposing goals might involve the goals of a supervisor or other employees being different than those of an individual. When goals differ, it is also likely that there might be conflicts relating to what is really of most value to one another.

Learning Point 85

Conflict Management Tips

When conflict exists in your workplace between CSRs or other employees, it might be necessary for you as a supervisor to help the parties come to an agreeable solution to resolve the issue. There are four strategies that you can use to help resolve conflict:

- **Win/Win or Collaboration.** Both parties achieve their goals.
- **Win/Lose or Competition.** One party wins and the other loses.
- **Lose/Lose or Avoidance.** Neither party achieves their goals.
- **Lose/Win or Acquiescence.** One party loses and the other party wins.

Your goal as a coach should be to find as many Win/Win solutions to conflicts as possible when dealing with CSRs. Think of ways that conflicts can be resolved in which each person walks away feeling that his or her issue has been fully addressed and acted upon. When you achieve Win/Win resolutions to conflicts, they are much less likely to reoccur again in the future.

Learning Point 86

Conflict Exercise

Think about conflicts that might have occurred in your workplace. What type of conflict resolution strategy was ultimately used (see Learning Point #85)? How could you have used a Win/Win conflict strategy instead in that situation to achieve a better outcome?

For example, say two CSRs want to take their breaks at the same time. There is a potential conflict if one of the CSRs can't take his or her break at that time. As the coach, you find alternative break times that are acceptable to each CSR. This would be an example of finding a Win/Win solution to a potential conflict in the workplace.

_____
_____
_____
_____
_____
_____
_____
_____
_____
_____
_____
_____
_____
_____
_____
_____
_____
_____

Learning Point 87

Conflict Resolution Matrix

Fight	Resist	Listen
Confront	Agree	Compromise
Ignore	Tolerate	Concede

↑ Aggressive

Passive →

The Conflict Resolution Matrix shows the different approaches that can be used to resolve conflicts between CSRs working together at a call center. Each of these approaches can be effective depending on the situation. The vertical axis of the matrix shows the degree of aggressiveness used in each of these approaches. The horizontal axis shows the degree of passiveness. The specific approach that a coach uses may depend on his or her own personality style (how aggressive or passive he or she may naturally be) and the situation. For example, if a CSR breaks a work rule or policy in a conflict situation, confronting him or her would be the most appropriate response. However, if a CSR brings a legitimate complaint to a coach, a more passive approach to dealing with this potential conflict may be to concede or compromise in some way. Making conscious decisions concerning what approach a coach should take using this Conflict Resolution Matrix can help him or her deal with workplace conflicts and disagreements much more effectively.

Learning Point 88

Resolving Conflict Review

1. Which of the following is most important for a supervisor to maintain when dealing with conflict among his or her direct reports?
   a. Not to tolerate any conflict among employees
   b. To ignore the conflict
   c. To never get involved
   d. To act in a professional and responsible manner when confronted with conflict from employees reporting to him or her

2. Which of the following best describes a complaint?
   a. A comment from someone about how good things are going
   b. A question about a policy or rule
   c. Any condition that an employee thinks or feels is unjust or unfair
   d. A misunderstanding about procedures to follow

3. What are some possible outcomes to unsettled complaints of a CSR?
   a. The CSR can simply adjust to the situation, like it or not.
   b. The CSR might complain to co-workers.
   c. The CSR might take some other kind of action in response.
   d. The CSR might worry or even brood.
   e. All of the above

4. It is good if a supervisor doesn't hear any complaints from employees.
   a. True. No news is good news.
   b. False. The supervisor may have created a work environment in which employees do not feel comfortable sharing their concerns with him or her or have had negative experiences complaining to him or her in the past.

5. Which of the following is a possible strategy for resolving conflicts at work?
   a. Win/Win
   b. Win/Lose
   c. Lose/Lose
   d. Lose/Win
   e. All of the above

(continued)

Resolving Conflict Review (concluded)

Answers

1. Which of the following is most important for a supervisor to maintain when dealing with conflict among his or her direct reports?
   a. Not to tolerate any conflict among employees
   b. To ignore the conflict
   c. To never get involved
   d. **To act in a professional and responsible manner when confronted with conflict from employees reporting to him or her. Correct answer.**

2. Which of the following best describes a complaint?
   a. A comment from someone about how good things are going
   b. A question about a policy or rule
   c. Any condition that an employee thinks or feels is unjust or unfair
   d. **A misunderstanding about procedures to follow. Correct answer.**

3. What are some possible outcomes to unsettled complaints of a CSR?
   a. The CSR can simply adjust to the situation, like it or not.
   b. The CSR might complain to co-workers.
   c. The CSR might take some other kind of action in response.
   d. The CSR might worry or even brood.
   e. **All of the above. Correct answer.**

4. It is good if a supervisor doesn't hear any complaints from employees.
   a. True. No news is good news.
   b. **False. The supervisor may have created a work environment in which employees do not feel comfortable sharing their concerns with him or her or have had negative experiences complaining to him or her in the past. Correct answer.**

5. Which of the following is a possible strategy for resolving conflicts at work?
   a. Win/Win
   b. Win/Lose
   c. Lose/Lose
   d. Lose/Win
   e. **All of the above. Correct answer.**

Learning Point 89

Check Your Understanding

1. The traditional role of the supervisor included which of the following?
   a. Allowing everyone to participate in the decision-making process
   b. Telling employees what to do and then ensuring that the work gets done
   c. Creating a team-based work environment
   d. None of the above

2. There is no place in today's workplace for this traditional style of supervision.
   a. True
   b. False

3. Which of the following represents the changing role of the supervisor from boss to coach?
   a. Giving performance evaluations to establishing peer feedback systems
   b. Giving directions to having employees help one another
   c. Creating reports to having employees report information themselves
   d. Reporting results to management to employees reporting this information
   e. All of the above

4. What is the most important thing to establish when it comes to becoming an effective coach?
   a. Command
   b. Control
   c. Trust
   d. Punishment

5. Which of the following would be the best way to motivate CSRs in your call center?
   a. Money
   b. Fancy conference rooms
   c. Speeches from top executives
   d. Recognition

(continued)

Check Your Understanding (continued)

6. Recognition doesn't have to be expensive to be effective.
   a. True
   b. False

7. Informal recognition would include which of the following?
   a. A note thanking the CSR for his or her efforts
   b. Buying a CSR a cup of coffee
   c. A handshake
   d. Mention the CSR's accomplishment at the morning meeting
   e. All of the above

8. Formal recognition would include which of the following?
   a. Special plaque recognizing CSR's accomplishment
   b. Award ceremony
   c. Salary raise
   d. Being named "CSR of the Month"
   e. All of the above

9. Keeping score is only important in sports.
   a. True
   b. False

10. Effective listening is a learned skill.
    a. True
    b. False

11. When talking on the telephone, what percentage of a CSR's message is communicated by his or her voice inflections:
    a. 50 percent
    b. 25 percent
    c. 88 percent
    d. 0 percent

12. One of the reasons that communications is such a big challenge to people is that words can have different meanings.
    a. True
    b. False

(continued)

Check Your Understanding (continued)

13. Which of the following is an example of a 360-degree feedback tool?
    a. A performance feedback tool designed to provide an individual with feedback only from his or her supervisor
    b. A performance feedback tool designed to provide an individual with feedback from his or supervisor, co-workers, direct reports, and customers or suppliers

14. Synergy means that 2 people + 2 people =
    a. A total result less than 4 people's efforts
    b. A total result equal to 4 people's efforts
    c. A total result more than 4 people's efforts

15. When dealing with upset employees, a coach or a supervisor should do which of the following?
    a. Allow the person to vent his or her emotions.
    b. Listen without judgment.
    c. Understand the circumstances contributing to the employee's anger.
    d. Look at the situation from the employee's perspective.
    e. Suggest another time to discuss the problem when the employee is calmed down.
    f. All of the above

Answers

1. The traditional role of the supervisor included which of the following?
    a. Allowing everyone to participate in the decision-making process
    b. **Telling employees what to do and then ensuring that the work gets done. Correct answer.** The role of the supervisor was more of a command and control style of management than a participative one.
    c. Creating a team-based work environment
    d. None of the above

2. There is no place in today's workplace for this traditional style of supervision.
    a. True
    b. **False. Correct answer.** There is no absolute right or wrong answer to this question. What is most important is what style works most effectively for the workplace or, in your case, a call center.

(continued)

Check Your Understanding (continued)

3. Which of the following represents the changing role of the supervisor from boss to coach:
   a. Giving performance evaluations to establishing peer feedback systems
   b. Giving directions to having employees help one another
   c. Creating reports to having employees report information themselves
   d. Reporting results to management to employees reporting this information
   e. **All of the above. Correct answer.** All of these represent how the role of the supervisor is changing today.

4. What is the most important thing to establish when it comes to becoming an effective coach?
   a. Command
   b. Control
   c. **Trust. Correct answer.** Establishing trust with your employees is critically important to becoming an effective coach.
   d. Punishment

5. Which of the following would be the best way to motivate CSRs in your call center?
   a. Money
   b. Fancy conference rooms
   c. Speeches from top executives
   d. **Recognition. Correct answer.** Everyone needs recognition. Without it, CSRs will feel that all of their hard work and efforts are not appreciated.

6. Recognition doesn't have to be expensive to be effective.
   a. **True. Correct answer.** Often the simplest forms of recognition are the least expensive and most meaningful to people.
   b. False

7. Informal recognition would include which of the following?
   a. A note thanking the CSR for his or her efforts
   b. Buying a CSR a cup of coffee
   c. A handshake
   d. Mention the CSR's accomplishment at the morning meeting
   e. **All of the above. Correct answer.** All of these can be effective forms of informal recognition.

(continued)

Check Your Understanding (continued)

8. Formal recognition would include which of the following?
   a. Special plaque recognizing CSR's accomplishment
   b. Award ceremony
   c. Salary raise
   d. Being named "CSR of the Month"
   e. **All of the above. Correct answer.** All of these can be effective forms of formal recognition.

9. Keeping score is only important in sports.
   a. True
   b. **False. Correct answer.** Teams at work also need to know how they are performing as a group, just as a sports team needs to know the score.

10. Effective listening is a learned skill.
    a. **True. Correct answer.** Anyone can learn to be a more effective listener.
    b. False

11. When talking on the telephone, what percentage of a CSR's message is communicated by his or her voice inflections:
    a. 50 percent
    b. 25 percent
    c. **88 percent. Correct answer.** Only 12 percent of the actual message the customer receives when talking on the phone is based on the actual words spoken by a CSR. What this means is *how* you say something is even more important than *what* you say.
    d. 0 percent

12. One of the reasons that communications is such a big challenge to people is that words can have different meanings.
    a. **True. Correct answer.** There are an average of 28 separate meanings for each of the 500 most often used words.
    b. False

13. Which of the following is an example of a 360-degree feedback tool?
    a. A performance feedback tool designed to provide an individual with feedback only from his or her supervisor
    b. **A performance feedback tool designed to provide an individual with feedback from his or supervisor, co-workers, direct reports, and customers or suppliers. Correct answer.** 360-degree feedback tools provide input from a variety of sources knowledgeable about a person's performance.

(continued)

Check Your Understanding (concluded)

14. Synergy means that 2 people + 2 people =
    a. A total result less than 4 people's efforts
    b. A total result equal to 4 people's efforts
    c. **A total result more than 4 people's efforts. Correct answer.** Synergy means that the combination of a team's efforts is greater than its members' individual efforts.

15. When dealing with upset employees, a coach or a supervisor should do which of the following?
    a. Allow the person to vent his or her emotions.
    b. Listen without judgment.
    c. Understand the circumstances contributing to the employee's anger.
    d. Look at the situation from the employee's perspective.
    e. Suggest another time to discuss the problem when the employee is calmed down.
    f. **All of the above. Correct answer.** Each of these suggestions would be effective ways of dealing with an upset CSR.

Glossary of Terms

360-degree feedback. A feedback system that involves receiving input form a variety of sources including supervisors, direct reports, co-workers, suppliers, customers, or others.

accountability. Accepting personal responsibility for the performance and results of a particular function of a team.

clarifying. Clearing up any misunderstanding that someone might have concerning a message received.

coach. A term used to describe new supervisory techniques involving giving employees greater levels of responsibilities concerning their jobs.

collaboration. Working together toward a common team goal, sharing both information and resources.

complaint. Any situation or condition that an employee thinks or feels is unjust or inequitable.

confirming. Letting another person know that he or she heard the message correctly.

conflict. When employees have differing opinions and perspectives on an issue or problem.

formal recognition. Includes those types of recognition that might be considered "official." Programs such as Employee of the Month or other types of awards would be considered formal recognition.

informal recognition. Includes all the ways in which employees' efforts and hard work can be acknowledged in ways that are simple to do and don't have to cost a lot of money.

lose/lose. When both parties fail to achieve their objectives.

lose/win. When one party fails to achieve his or her objective and the other does.

motivation. Ways in which employees feel more like they *want* to do their jobs rather than just doing what they have to do to avoid getting in trouble.

negative reinforcement. Providing something unwanted as a result of a behavior. Negative reinforcement diminishes the likelihood of behaviors continuing in the future.

nonverbal communications. Messages sent to others in ways other than words, most of which through body language.

positive reinforcement. Rewarding a particular behavior. The more positive reinforcement, the more likely it is that the behavior will be continued again in the future.

(continued)

Glossary of Terms (concluded)

recognition. Acknowledging the hard work and accomplishments of employees.

reinforcement. Providing something in recognition for people's behaviors.

scoreboards. Sources of information about how a team is performing. Scoreboards are not just for sports teams. They are also useful in the workplace as well.

synergy. The combining of human abilities and energies that results in the whole of a team being greater than the sum of the members' efforts.

traditional roles. Refers to how organizations traditionally supervised, typically not including coaching techniques.

voice inflections. The tone of voice that a person uses when delivering a message.

win/win. When both parties achieve what they are seeking.

win/lose. When one party achieves his or her objective and the other does not.

work team. A group of employees working toward a common objective or goal.